Cambridge Elements ≡

Elements of Paleontology

INCORPORATING RESEARCH INTO UNDERGRADUATE PALEONTOLOGY COURSES

or a Tale of 23,276 Mulinia

Patricia H. Kelley

University of North Carolina Wilmington

Paleontological
S O C I E T Y

CAMBRIDGE
UNIVERSITY PRESS

CAMBRIDGE
UNIVERSITY PRESS

University Printing House, Cambridge CB2 8BS, United Kingdom

One Liberty Plaza, 20th Floor, New York, NY 10006, USA

477 Williamstown Road, Port Melbourne, VIC 3207, Australia

314–321, 3rd Floor, Plot 3, Splendor Forum, Jasola District Centre, New Delhi – 110025, India

79 Anson Road, #06–04/06, Singapore 079906

Cambridge University Press is part of the University of Cambridge.

It furthers the University's mission by disseminating knowledge in the pursuit of education, learning, and research at the highest international levels of excellence.

www.cambridge.org
Information on this title: www.cambridge.org/9781108717892
DOI: 10.1017/9781108681643

First published 2018

A catalogue record for this publication is available from the British Library.

ISBN 978-1-108-71789-2 Paperback
ISSN 2517-780X (online)
ISSN 2517-7796 (print)

Incorporating Research into Undergraduate Paleontology Courses

or a Tale of 23,276 *Mulinia*

Elements of Paleontology

DOI: 10.1017/9781108681643
First published online: October 2018

Patricia H. Kelley
University of North Carolina Wilmington

Abstract: Research-led, research-oriented, and research-based teaching incorporate research into teaching to different degrees. Research-led teaching focuses on content and informs students about current research findings. Research-oriented teaching focuses on techniques and often occurs in research methods courses. In research-based teaching, students participate in research. Through this involvement, they benefit from improved content knowledge, research skills, and life skills, as well as enhanced personal development. Research-embedded courses can make such benefits available to a wide range of students. Best practices in experiential learning and the incorporation of research in teaching include intentionality, planning, authenticity, reflection, training, monitoring, assessment, and acknowledgment. These principles of best practice are illustrated by courses with embedded student research. Guidelines are presented for how to plan and execute a semester-long course-embedded research project, as well as alternative and shorter-term approaches. Research-based teaching provides challenges for students and faculty, but the benefits for all stakeholders are strong.

Keywords: research-based teaching, experiential learning, research-embedded courses, CURE, intentionality, reflection

ISBNs: 9781108717892 (PB), 9781108681643 (OC)
ISSNs: 2517-780X (online) 2517–7796 (print)

Contents

1 Introduction

> Everybody says I should be incorporating research in my undergraduate courses, and it would sure help my Broader Impacts section on my next NSF proposal. But I don't know how to do it! And even if I did know, how would I find the time? I'm too busy DOING research to spend time redesigning my courses to incorporate research! HELP! – *Anonymous Paleontologist*

Sound familiar? Perhaps you have heard – or made, or at least thought – such a comment. Let's examine the statements made by Anonymous Paleontologist (henceforth AP) more closely.

Is AP correct that paleontologists should be incorporating research in their undergraduate teaching? Traditionally teaching and research have been viewed as competing interests (Lopatto, 2010; Carleton University, 2017) or at best disconnected activities (Jenkins, 2001). This viewpoint is echoed by AP, who feels pressured by the various demands inherent in the academic profession. AP's viewpoint reflects the fact that academic units often treat research and teaching as unrelated activities (and evaluate faculty separately in each category; Brew, 2010), with one or the other valued more highly depending on the institution (Jenkins and Healey, 2012).

In recent years, however, there has been a push to integrate research and teaching. At the U.S. National Science Foundation (NSF), this push is reflected in the NSF strategic objective to "integrate education and research to support development of a diverse STEM workforce with cutting-edge capabilities" (National Science Foundation, 2014, p. 8). Activities that support this objective fall under the domain of "Broader Impacts" mentioned by AP; all NSF research proposals must address the broader impacts of the research to society and particular desired societal outcomes. One criterion initially proposed for merit review of broader impacts was "How well does the activity advance discovery and understanding while promoting teaching, training, and learning?" (National Science Board, 2011, p. 4). Such integration continues to contribute positively to a proposal's review score in the broader impacts category (personal observation, based on participation in the 2017 NSF Division of Earth Sciences Committee of Visitors; National Science Foundation, 2017).

But why this emphasis on integrating research and teaching? The 2014–2018 NSF Strategic Plan (National Science Foundation, 2014, p. 8) states:

> One of NSF's most enduring contributions to the national innovation ecosystem is the integration of education and research in the activities we support. When students participate in cutting-edge research activities under

the guidance of the Nation's most creative scientists and engineers, the students can gain the up-to-date knowledge and practical, hands-on experience needed to develop into creative contributors who can engage in innovative activities throughout all sectors of society.

Other benefits accrue as well from incorporating research into teaching (Jenkins, 2001; Brew, 2010; Lopatto, 2010; Science Education Resource Center, 2016). Students become more engaged in learning and more motivated to pursue science. They understand scientific reasoning better and are poised to be better citizens (Carleton University, 2017) who are prepared to evaluate the validity of competing claims and make evidence-based decisions (in daily life and in the voting booth!) – a benefit of vital importance to the United States, where much of the population is scientifically illiterate (Lopatto, 2010) and even downright hostile to science. In a world in which factual knowledge (as well as misinformation) can be accessed on smart phones (see Brew, 2010), a premium is placed not on delivering facts but on promoting students' abilities to "create and find and synthesize new knowledge" (Jenkins, 2001, p. 18). Development of critical thinking and problem-solving skills, which are fostered by integration of research and teaching (Lopatto, 2010), will be essential for success in what has been referred to as the "knowledge economy." Brew (2010, p. 141) concluded:

> Students are going to need to be able to critically evaluate knowledge; to make rational judgments in the light of good evidence, evidence that they perhaps gather, and to reflect on what they are doing and why. These are the skills of critical inquiry, which are central to a super-complex society. Today's society demands creativity. It demands the ability to deal with complexity and uncertainty. We need new kinds of teaching, new spaces, new ideas about knowledge, new ways to engage students. I believe that the integration of research and teaching provides exciting ways to meet this agenda.

But will it help students learn paleontology? If "learning paleontology" means knowing the stratigraphic ranges of strophomenid brachiopods, perhaps not, unless students are specifically involved in research that includes strophomenids. But if "learning paleontology" means being able to apply paleontological concepts discussed in class, my experience indicates that incorporation of research into teaching does lead to improved learning of paleontology (see student reflections below). After all, students can look up information about strophomenid brachiopods on their smart phones, but using that information to answer research questions or test hypotheses about evolutionary or ecological processes requires a deeper understanding of paleontology, which can be derived from research involvement.

1 Introduction

> Everybody says I should be incorporating research in my undergraduate
> courses, and it would sure help my Broader Impacts section on my next NSF
> proposal. But I don't know how to do it! And even if I did know, how would
> I find the time? I'm too busy DOING research to spend time redesigning my
> courses to incorporate research! HELP! – *Anonymous Paleontologist*

Sound familiar? Perhaps you have heard – or made, or at least thought – such a comment. Let's examine the statements made by Anonymous Paleontologist (henceforth AP) more closely.

Is AP correct that paleontologists should be incorporating research in their undergraduate teaching? Traditionally teaching and research have been viewed as competing interests (Lopatto, 2010; Carleton University, 2017) or at best disconnected activities (Jenkins, 2001). This viewpoint is echoed by AP, who feels pressured by the various demands inherent in the academic profession. AP's viewpoint reflects the fact that academic units often treat research and teaching as unrelated activities (and evaluate faculty separately in each category; Brew, 2010), with one or the other valued more highly depending on the institution (Jenkins and Healey, 2012).

In recent years, however, there has been a push to integrate research and teaching. At the U.S. National Science Foundation (NSF), this push is reflected in the NSF strategic objective to "integrate education and research to support development of a diverse STEM workforce with cutting-edge capabilities" (National Science Foundation, 2014, p. 8). Activities that support this objective fall under the domain of "Broader Impacts" mentioned by AP; all NSF research proposals must address the broader impacts of the research to society and particular desired societal outcomes. One criterion initially proposed for merit review of broader impacts was "How well does the activity advance discovery and understanding while promoting teaching, training, and learning?" (National Science Board, 2011, p. 4). Such integration continues to contribute positively to a proposal's review score in the broader impacts category (personal observation, based on participation in the 2017 NSF Division of Earth Sciences Committee of Visitors; National Science Foundation, 2017).

But why this emphasis on integrating research and teaching? The 2014–2018 NSF Strategic Plan (National Science Foundation, 2014, p. 8) states:

> One of NSF's most enduring contributions to the national innovation
> ecosystem is the integration of education and research in the activities we
> support. When students participate in cutting-edge research activities under

the guidance of the Nation's most creative scientists and engineers, the students can gain the up-to-date knowledge and practical, hands-on experience needed to develop into creative contributors who can engage in innovative activities throughout all sectors of society.

Other benefits accrue as well from incorporating research into teaching (Jenkins, 2001; Brew, 2010; Lopatto, 2010; Science Education Resource Center, 2016). Students become more engaged in learning and more motivated to pursue science. They understand scientific reasoning better and are poised to be better citizens (Carleton University, 2017) who are prepared to evaluate the validity of competing claims and make evidence-based decisions (in daily life and in the voting booth!) – a benefit of vital importance to the United States, where much of the population is scientifically illiterate (Lopatto, 2010) and even downright hostile to science. In a world in which factual knowledge (as well as misinformation) can be accessed on smart phones (see Brew, 2010), a premium is placed not on delivering facts but on promoting students' abilities to "create and find and synthesize new knowledge" (Jenkins, 2001, p. 18). Development of critical thinking and problem-solving skills, which are fostered by integration of research and teaching (Lopatto, 2010), will be essential for success in what has been referred to as the "knowledge economy." Brew (2010, p. 141) concluded:

> Students are going to need to be able to critically evaluate knowledge; to make rational judgments in the light of good evidence, evidence that they perhaps gather, and to reflect on what they are doing and why. These are the skills of critical inquiry, which are central to a super-complex society. Today's society demands creativity. It demands the ability to deal with complexity and uncertainty. We need new kinds of teaching, new spaces, new ideas about knowledge, new ways to engage students. I believe that the integration of research and teaching provides exciting ways to meet this agenda.

But will it help students learn paleontology? If "learning paleontology" means knowing the stratigraphic ranges of strophomenid brachiopods, perhaps not, unless students are specifically involved in research that includes strophomenids. But if "learning paleontology" means being able to apply paleontological concepts discussed in class, my experience indicates that incorporation of research into teaching does lead to improved learning of paleontology (see student reflections below). After all, students can look up information about strophomenid brachiopods on their smart phones, but using that information to answer research questions or test hypotheses about evolutionary or ecological processes requires a deeper understanding of paleontology, which can be derived from research involvement.

understanding of content through in-class activities and interactions with one another.

In keeping with best practices, RLT need not be restricted to transmission of research results via lecture (which Griffiths, 2004, referred to as research "weakly embedded" in teaching). In contrast, "strongly integrated" research is "used deliberately to shape the learning activities carried out by students" (Griffiths 2004, p. 721). Various activities that more strongly integrate research and teaching are available, e.g., assigning journal articles for reading and class discussion (Darden, 2003). Robinson (1987) advocated this approach in introductory geology classes, noting that success depends on the topic chosen and the preparation students receive for each assignment. To help introductory students handle complex topics, Klemm (2013) used Adapted Published Research Reports, re-writes of published articles to make them more accessible to first-year students; students worked independently and then in teams to answer a series of questions resembling those asked of reviewers of journal submissions. He reported that, in addition to learning content, students gained understanding of the research process, including scientific reasoning and argumentation, and developed critical thinking and analytical skills in this "minds-on" approach. Activities in which students conduct literature searches, develop annotated bibliographies, and write or critique research proposals (Peterson et al., 1996; Darden, 2003; Science Education Resource Center, 2016) may develop similar skills. Such approaches are usable by faculty who are research active as well as those not currently engaged in research projects. And faculty may benefit from these activities as well, e.g., if students discover articles relevant to their instructor's research (Darden, 2003) – mitigating AP's "I'm so busy doing research that I don't have time for this" complaint.

Research-led teaching, beyond simple conveyance of information, may also include laboratory or other exercises using real data to provide a "minds-on" approach that also includes aspects of ROT. For example, I have provided students in historical geology and invertebrate paleontology with data I collected to test punctuated equilibrium (Kelley, 1979, 1984). Students graphed and analyzed the data and interpreted what mode of evolution occurred – and realized that interpretations are not always straightforward. The online availability of real datasets has increased the ease of incorporating such an approach (Wei and Woodin, 2011). For example, Gutiérrez and Baker (2013) described a class exercise in which students analyzed online soil data using methods found in the literature; different students selected different methods to use and collaborated to compare their results. Use of real data in RLT helps students realize that science is messy (Gutiérrez and Baker, 2013), in contrast to "cookbook" experiments that always turn out "right" if students follow the instructions. Ellwein et al. (2014) found similar results from surveying students who used long-term climate science data sets. Working with authentic data serves to engage students (although not as much as hands-on data collection; Gold et al., 2015). Students develop higher-order thinking skills and understand the scientific process better.

2.2 Research-Oriented Teaching

Research-oriented teaching, as defined by Griffiths (2004), focuses on how knowledge is produced in the discipline, rather than on learning that knowledge itself. ROT is exemplified by methods courses, which are designed to train students in the techniques of the discipline. Such courses may provide instruction and experience in field methods, use of instrumentation and analytical methods, geospatial and quantitative methods, and use of programming languages and software environments such as R and MATLAB. Methods courses are common within geoscience curricula. For instance, at University of North Carolina Wilmington (UNCW) the "Field Methods in Geosciences" course is described as "Introduction to methods and techniques used in the geosciences including field measurement, sample retrieval and data analysis" (University of North Carolina Wilmington, 2017). For BS Geology students, this course is followed by "Techniques in Applied Geology" and ultimately the "Field Course in Geology," a traditional field camp-style course. Our department also offers techniques-based courses in quantitative methods, oceanography, Geographic Information Systems, remote sensing, and cartography, as well as special topics methods courses (e.g., petrographic techniques).

Although research methods courses need not engage students in actual research (National Academies, 2017), in some ROT students may be working with existing data sets or collecting real data in a manner similar to RBT. For instance, Hopper et al. (2013) described a 1-credit-hour research course in which students experienced authentic research tasks of collecting and analyzing Doppler radar data. Koretsky et al. (2012) described a field-based environmental geochemistry course in which students learned field and laboratory research skills to investigate water quality in a local lake, a project they characterized as "authentic inquiry." Such inquiry shares some of the characteristics of research (Healey and Jenkins, 2017) and yields a variety of benefits. Inquiry-based teaching improves learning of content, develops student skills in problem solving and critical thinking, and allows students to "practice the activities involved in science" (Apedoe et al., 2006, p. 414; see recommendations in Apedoe et al., 2006 for how to implement inquiry in undergraduate geoscience courses).

2.3 Research-Based Teaching

In inquiry, the results of an investigation are unknown to the student but are not new to the scientific community (and probably not to the instructor); in contrast, research yields information new to the scientific community (Lopatto, 2010; Auchincloss et al., 2014). Therefore research-based teaching goes beyond the approach of inquiry to involve students in research experiences that generate new knowledge, i.e., "an original intellectual or creative contribution to the discipline" (Wenzel, 1997, p. 163). Much has been written about the benefits of research experiences for undergraduates (see summaries in, e.g., Lopatto 2010; Kelley and Visaggi, 2012; Koretsky et al., 2012; Corwin et al., 2015; Kortz and Kraft, 2016; National Academies, 2017). Benefits extend beyond enhancement of scientific skills (reading literature, developing and testing hypotheses, analyzing data) and life skills (problem solving, critical thinking, communication) to personal development (self-confidence, ability to work independently and in teams).

At many institutions, these benefits are reserved for a select group of students (Jenkins, 2001), determined by such criteria as grade point average or acceptance into Honors programs. Typically, such experiences follow an apprenticeship model: members of this privileged group work one-on-one with a faculty mentor on a research project or thesis, or they enroll in a research internship or co-op experience. These capstone experiences involve a significant investment by the research supervisor, who likely also has research deadlines and goals to meet (per AP's concern). Consequently, the faculty mentor may not wish to expend effort on less promising or less motivated students (Jenkins, 2001; see also Kortz and Kraft, 2016). Thus there has been a tendency to resist offering research opportunities to the entire student body. But if RBT yields such benefits, shouldn't all students have the opportunity to participate? I agree with Healey and Jenkins (2009, p. 3; see also Healey and Jenkins, 2017) that "*all* undergraduate students in *all* higher education institutions should experience learning through, and about, research and inquiry."

An effective avenue for broadening the participation of students in research is through research-embedded courses, referred to in some cases as CUREs (Course-based Undergraduate Research Experiences; Auchincloss et al., 2014). Such courses provide research opportunities for a wider range of students (e.g., students with less experience or less stellar academic records) and confer benefits comparable to those of the apprenticeship model (Lopatto, 2010; Kelley and Visaggi, 2012; Auchincloss et al., 2014; Corwin et al., 2015; National Academies, 2017). According to Auchincloss et al. (2014), CUREs involve students in the activities of science (hypothesis development, study design, data collection and analysis, interpretation, dissemination); result in discovery of new knowledge that is relevant to the discipline; and are iterative (see also American Library Association, 2015), building on previous knowledge (e.g., findings of previous students in the course). CUREs also involve collaboration among participants, an advantage not usually offered by the apprenticeship model (see Burke, 2011, for a discussion of benefits and best practice in employing group work). Research-embedded courses are becoming more common in the sciences (National Academies, 2017), with good examples from the geosciences provided by Foos (1997; geochemistry lab course), Mayborn and Lesher (2000; advanced igneous petrology course), Gonzales and Semken (2006; field-based igneous petrology course), Davies-Vollum (2006; sedimentology course), and Montgomery and Donaldson (2014; introductory honors paleontology course). At some institutions, the curriculum is structured to include a sequence of CUREs that provide students with a multi-year sustained research experience (e.g., Allen et al., 2017).

In the next section of the Element I focus on a semester-long research-embedded course in invertebrate paleontology (IP) I taught for a dozen years at UNCW. This course provides an opportunity to illustrate best practices in experiential learning and the incorporation of research in undergraduate teaching. The structure of the course is described briefly here, with more details presented by Kelley and Visaggi (2012). Figure 1 provides a general schedule for student activities and deliverables, as well as a timeline for instructor implementation of best practices in experiential learning.

IP is a four-credit-hour elective course with three hours of lecture and three hours of laboratory each week. As I taught it, the content combined paleontological principles with discussion of taxonomic groups. The laboratory provided hands-on work each week in a traditional format (observing, sketching, and answering questions about specimens from the teaching collection) related to the lecture topics, after which students spent approximately two hours working on a team research project worth 20 percent of the entire course grade. Two or three teams (depending on class size, which usually ranged from 10 to 14 and averaged 11 students) were each assigned a bulk sample from a Cenozoic site in the region, often one they collected themselves on a field trip early in the semester (Figure 2) but in some cases an archived but unprocessed sample. Students wet-sieved the samples to remove the specimens from their matrix, sorted them into species, identified them at least to genus level, developed hypotheses, and collected data on abundance, predation traces, life modes, and/or morphology depending on the focus of the project that semester. Each student produced a paper coauthored with team members in professional

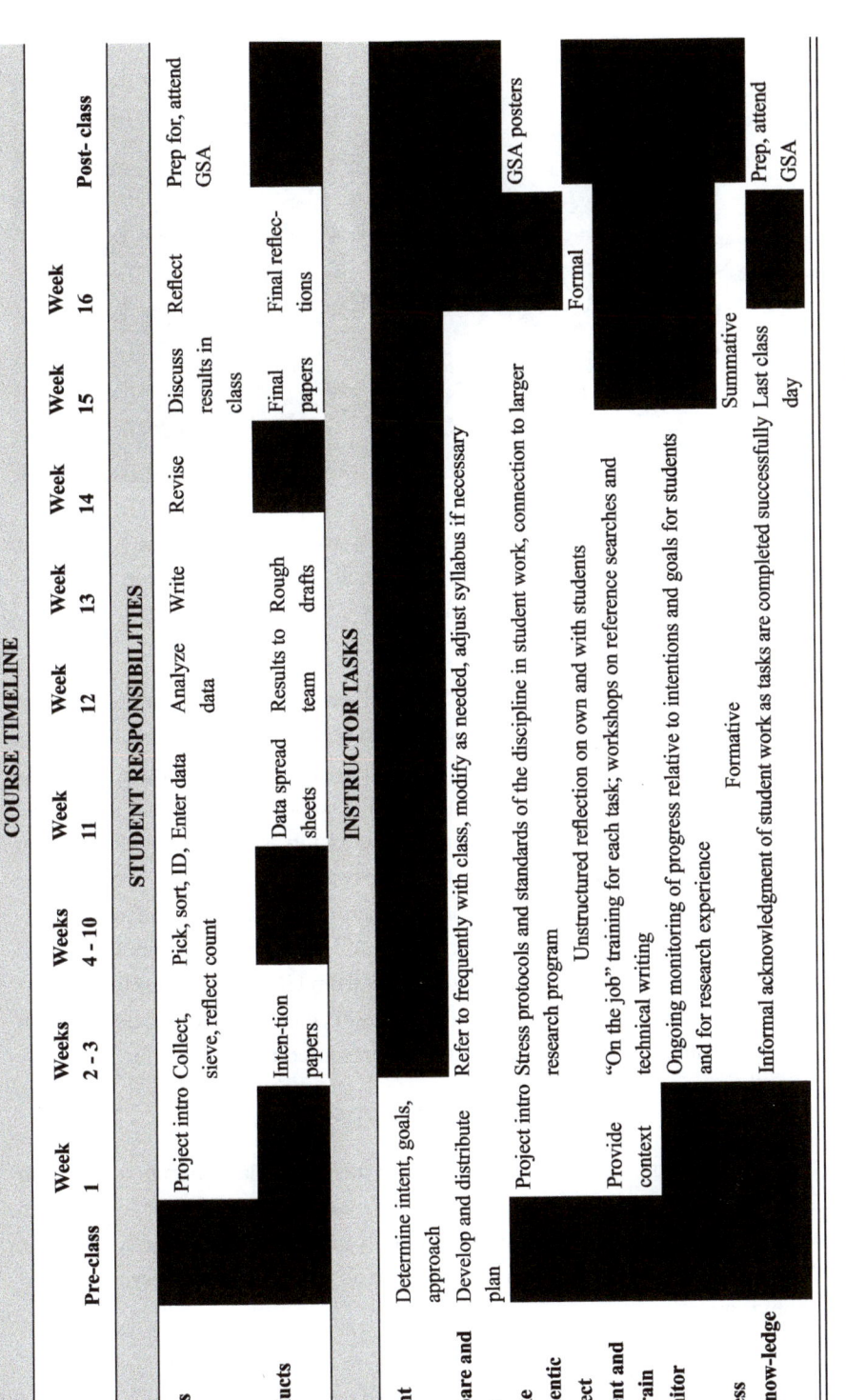

Figure 1 General timeline for the Invertebrate Paleontology course, showing weekly student responsibilities (tasks and products) and the timeline of instructor implementation of eight principles of best practices in experiential learning.

Figure 2 Invertebrate paleontology students wet sieving samples in the field collected at Kirby Pond, Timmonsville, South Carolina, for use in the 2015 team research project. Photo by Chetara Davis King.

journal format, with individually written sections and contributions from other group members. Finally, we submitted abstracts synthesizing the results produced by the different teams to a Geological Society of America (GSA) meeting. Because abstracts were due after the semester ended, and all student data required vetting prior to abstract submission, students usually did not participate in writing the formal GSA abstracts. However, I urged all IP students to present at GSA the following semester; students who were interested (and who could balance work, family, or other commitments in order to attend) applied for and received travel grants from UNCW's Center for Support of Undergraduate Research and Fellowships. Typically, several students from each class attended the GSA meeting and presented the posters.

3 A Tale of 23,276 *Mulinia*

Let's listen in on the IP laboratory on a typical Tuesday afternoon in spring 2013. The students and I were pouring over a vast array of mollusc fossils from the Plio-Pleistocene Waccamaw Formation of Horry County, South Carolina. Note that I have substituted names of family members to preserve students' anonymity.

"I hate *Mulinia*!" Katherine declared.

McKenzie commiserated. "I know – I've counted over a thousand *Mulinia* this afternoon!"

"Me too. At least I think they're *Mulinia*. Some of them might be *Spisula*," Owen sighed.

"If you want to ID these last gastropods, I'll switch to *Mulinia*," Annie offered, "if that's ok." She looked at me.

"You guys can organize your team however you want in order to get everything done," I responded. "Remember, you need to finish species identifications this week so I can check them. There are only a few weeks left to tabulate and analyze data and test your hypotheses. Plus we need to make sure there's time to share rough drafts of your sections with your teammates and me so you can make revisions."

I spoke confidently, but underneath I was a bundle of nerves. Since 2003, I had included a semester-long authentic, publishable research project in IP. There was always pressure to produce publishable data, but this year it was worse. For our regional accreditation, UNCW was embarking on a Quality Enhancement Plan (QEP), ETEAL ("Experiencing Transformative Education through Applied Learning" – a nice reference to our school color; ETEAL, 2018). I had volunteered Invertebrate Paleontology as an ETEAL early adopter class, and I knew I had some work to do on my RBT pedagogy. While participating in UNCW's QEP institute on applied learning the previous summer, I realized that my course was missing pieces of intentionality and critical reflection. I knew the class would figure prominently in QEP assessment and that my colleagues at UNCW would be watching closely. I had to make it work!

But the samples weren't cooperating! I was alarmed when we sieved our bags of sediment and found the most discouraging fossil assemblage in 20 years of incorporating authentic research in graduate and undergraduate classes. Among more than 60 taxa to be identified were thousands of *Mulinia lateralis* – a nondescript species of bivalve a few millimeters in size, and way too similar to *Spisula*!

"Let's have a contest to guess the number of *Mulinia*," John laughed. "Winner gets an A!"

"I'll take the winner to Denver to present the projects," I said.

"Ha! You know we're going regardless!" Timothy retorted. And the students laughed and bent over the samples with renewed enthusiasm. Ultimately, despite identifying and counting 23,276 *Mulinia*, their final critical reflection pieces were glowing, and a majority of the class presented their results at the national GSA meeting in Denver (Kelley et al. 2013a, b).

3.1 Best Practices for Research-Embedded Courses

Let's examine the best practices for a research-embedded course illustrated by this vignette. Why was this research experience successful despite the challenges of the materials being studied? A useful framework for understanding the course's successes and challenges is provided by the National Society for Experiential Education (NSEE, 1998) – the eight "principles of best practice in experiential education." Although these practices are applicable for pedagogies other than RBT, they provide appropriate guidelines for how to plan and execute a course-embedded research project (see Figure 1).

3.1.1 Intention

"Intention represents the purposefulness that enables experience to become knowledge" (NSEE, 1998). In this first principle, NSEE (1998) stressed that all parties (instructor and students) involved in the experience need to understand why it was selected as the mode of learning. Prior to embarking on the IP 2013 experience, I thought through all aspects of the experience, including what I wanted the students to do, why I wanted them to do it, and what I hoped they (and I) would get out of it. I then created a detailed handout (provided in Appendix 1) that included the rationale and instructions for the project. It presented the scientific context and purpose of the study: Why are we doing this? What important scientific questions can we answer? What previous work has been done and what questions are left unanswered (see Auchincloss et al., 2014, and American Library Association, 2015)? The handout also included the university-wide ETEAL Student Learning Objectives (SLOs), the first of which was to "articulate their expectations, the purpose, and/or the goals of the experience in terms of their personal educational development." Students completed an initial reflection piece addressing this SLO, which made them a party to the intention component.

In this initial intention piece, some students expressed an expectation regarding content knowledge. For example, one student stated, "I believe this experience will shed some light on certain aspects of paleontology, geology, and oceanography that I do not understand as well as I would like."

Others looked forward to understanding the research process, as expressed by this student: "I believe … that I will gain experience and knowledge in how a research project is supposed to progress and how to make sure that when I begin a research project of my own, I will know how to ensure that both my process and research are done properly." Another student anticipated experiencing collaboration: "I also expect that I will gain valuable experience in how to work on a research project with others, since this will be a group effort."

Other students focused on preparation for a career:

> I believe that this will enhance my ability to think critically and translate my learning into a written format. In the information age it is important that

students have this ability and gain a competitive edge in the job market . . ..
Learning to use information is critical in understanding and analyzing objectives. Much of this course is dedicated to analyzing data. I consider this to be
key to my development as a scientist as analyzing and forming conclusions is
fundamental. The purpose of this course is to provide an applied learning
experience that has a very "hands-on" component. I expect to strengthen my
abilities as a critical thinker, writer, and member of the scientific community.

> Normally lab will consist of a lecture filled with pages of notes followed
> by a lab which also consists of many pages of questions both fill in the black
> [sic] and short answer. This lab however was very different . . . most of the
> lab consisted of applied hands on learning. It was a great wa[y] to forge new
> bonds with classmates and get our hands dirty . . .. Gaining knowledge of
> paleontological practices would be a great skill to add to my repertoire.

I was impressed with how much the intention component benefited the research
experience. Because students thought specifically about what they hoped to
gain from the ETEAL project, they seemed more eager than in previous years to
finish the "traditional" part of the lab each week and to jump into sorting and
identifying their specimens.

3.1.2 Preparedness and Planning

Without adequate preparation of students to undertake authentic research,
the experience will not be successful. Such preparation requires careful
planning (Healey and Jenkins, 2017). NSEE (1998, p. 1) recommended
that such plans should be referred to regularly but "be flexible enough to
allow for adaptations as the experience unfolds." Kortz and Kraft (2016)
also discussed the importance of providing detailed directions to students in
a course-embedded research project, along with the need for flexibility.
To prepare the students for the experience, I provided clear guidelines in
the initial project handout (Appendix 1), which included the geological
background, the SLOs, detailed instructions for completing the project,
instructions for the final critical reflection, a timeline for completing the
study (see Kortz and Kraft, 2016), and some references to get the students
started on their research. We discussed the information in detail at the
beginning of the semester and referred to it frequently as the project pro-
gressed, which enabled the work to go more smoothly. In addition, I enlisted
UNCW's Science Librarian and the Writing Services Coordinator to provide
workshops training students in reference searches and technical writing.
Also at the beginning of the semester, I provided the rubric that would be
used for grading the final research paper (see Appendix I of Kelley and
Visaggi, 2012), so students would know how to structure their final product.

These approaches are consistent with the recommendation of Healey and Jenkins (2017) that good mentoring of student research includes guidance on how students can share their results effectively.

Despite the considerable planning, unanticipated challenges arose, requiring flexibility, as noted by NSEE (1998). Time constraints are a significant limitation in course-embedded research experiences (see also Davies-Vollum, 2006). The time constraints were especially acute this semester; most samples that I have used in my teaching contain perhaps hundreds to a few thousand specimens, not the approximately 25,000 specimens found in 2013. However, once we had wet-sieved the samples and begun to work on them we could not cut corners (i.e., only examine a portion of the specimens) because the results would not be robust scientifically. Students had to spend much more time sorting, identifying, and counting specimens than I had anticipated. Consequently, I had to adjust the syllabus to give them more time to complete this work, so that they would have adequate opportunity to analyze and interpret their data and write up their results. For instance, I switched some lecture periods to lab periods so that students could finish their counts in a timely manner, and then we made up the lecture time during a subsequent lab time after the data tabulations were complete.

3.1.3 Authenticity

Authenticity provides a more compelling learning experience (Koretsky et al., 2012; Ellwein et al., 2014; Gold et al., 2015; but see National Academies, 2017, for comments on the utility of the "authentic" or "nonauthentic" designation). I made students aware that the project represented authentic research that used protocols consistent with the standards of the discipline. This particular year, the research project extended work done during an NSF-sponsored Research Experiences for Undergraduates (REU) program that I directed from 2008 to 2013 (Kelley and Dietl, 2012). Students recognized that they were participating in a much larger research program (i.e., work was iterative, as promoted by, e.g., Auchincloss et al., 2014, the American Library Association, 2015, and the National Academies, 2017) and were excited to contribute their "piece of the puzzle." Students were made to feel a part of the scientific community by knowing that they would be coauthors on a peer-reviewed abstract and would have an opportunity to present their work at a national meeting (see Corwin et al., 2015, for discussion of such external validation). This expectation served to motivate them; one student commented in his intention piece, "I can't help but be excited for an opportunity to work on a group project that has the potential to be published." They also knew that their samples would be reposited at the Paleontological Research Institution in Ithaca, New York, upon completion of the study, per the REU grant. Thus, they recognized that their research needed to be of the highest quality, and they worked hard to produce accurate results. Students internalized the process of conducting research in paleontology from conception to dissemination – a hallmark of authenticity.

3.1.4 Reflection

Understanding is enhanced when students are encouraged to reflect about both the scientific study under investigation and the work they are conducting as part of the study (National Academies, 2017). According to the NSEE (1998, p. 1), reflection "transforms simple experience to a learning experience … from identifying intention … to considering preconceptions and observing how they change as the experience unfolds. Reflection is also an essential tool for adjusting the experience and measuring outcomes." Kortz and Kraft (2016) also stressed the role of reflection in allowing students to internalize the research process in which they are involved and in allowing the instructor to improve the learning experience.

Prior to 2013 the course had not included a reflection component, and both the students and I felt some discomfort in writing our reflection pieces; we were all a little concerned whether we were doing it "right." I built two formal reflection opportunities into the students' experience. At the beginning of the experience, students composed an intention piece as described above. Following the experience, they completed a final reflection piece that considered both the academic enhancement and personal growth that resulted from the work, using the prompts shown in Appendix 1. For instance, students were asked to reflect on how the experience helped them understand concepts in paleontology and on the academic and personal skills they used during the experience. They were also asked to reflect upon how the research experience facilitated their learning, upon how they and their group performed, and upon the implications of what they learned from the experience for their future. Comments from these reflection pieces are included below.

Unstructured reflection opportunities occurred during the weekly lab sessions as students discussed their progress as they worked on the samples. Students also used the final day of class, when the teams presented their work to one another, to reflect on their experience and discuss their results.

I also put my own reflections in writing. My own evaluation of the experiential process emphasized the need for flexibility and development of alternative approaches. This recognition was useful in planning the next iteration of the course two years later (spring 2015), when I rethought the timeline for various activities. For instance, I asked one of my graduate students to visit a locality near his home in South Carolina, where he collected bulk samples prior to the start of the spring semester. Students could thus begin working on those samples early in the semester, rather than wait to start the project in late February, when the class sampled that locality and another.

3.1.5 Orientation and Training

NSEE (1998, p. 1) recommended that "ongoing structured development opportunities should also be included to expand the learner's appreciation of the context and skill requirements of her/his work." Context was provided on the first day of class, when we discussed the designation of this course as an ETEAL course and what was meant by experiential/applied learning. The initial handout and our discussion at the beginning of the semester served to orient students to applied learning and to the purpose of ETEAL. The handout also served as an orientation to the scientific problem being studied in the applied learning experience (Appendix 1).

Training was "on the job" – as each new task was begun (sieving, picking, sorting, identifying, counting, data tabulation), I worked with each team to make sure that they were properly trained in the necessary techniques. Students negotiated within their teams to determine who would be responsible for each type of data analysis (such specified responsibilities promote team member accountability and effort; Burke, 2011). I trained each student individually in the procedures needed for his or her component of the analysis (e.g., diversity, life mode, predation).

3.1.6 Monitoring and Continuous Improvement

To maximize a learning experience, formative evaluation should occur by continuous monitoring of progress relative to intentions and goals – both for individual students and for the research experience itself (NSEE, 1998). The ETEAL initiative funded one of my graduate students, who was experienced in working with bulk samples, to contribute four hours per week to the 2013 project. She supervised one team of students while I supervised the other. We worked alongside the students on the samples during each lab, constantly monitoring their progress. We made sure appropriate research protocols were followed and we checked student work for accuracy as it was completed. For instance, we meticulously went through each sample after the students completed picking it for usable specimens, and we extracted any suitable specimens the students had overlooked. Students emailed their data spreadsheets to me for vetting before I distributed them to other members of the team. Following data analysis, we went over the spreadsheets together in lab, e.g., to make sure that their calculations of predation metrics were correct, or their rarefaction analyses were being conducted properly, before sending results to teammates. They also submitted drafts of the report sections to me for critique prior to distribution to other team members. Students even had opportunity to modify their manuscripts after the final day of discussion based on oral feedback received that day from their classmates and me (see also Mayborn and Lesher, 2000, for a similar approach). Students therefore received continuous feedback on their work and could adjust and improve as needed. Kortz and Kraft (2016) also found frequent feedback to students to be essential to the success of their research projects. In addition, continuous monitoring allowed me to modify the learning experience as needed by adjusting scheduling (see above).

3.1.7 Assessment and Evaluation

NSEE (1998) stressed the importance of assessment of progress towards the specific learning goals and objectives (and modification of those goals if needed) and evaluation of the experiential process regarding whether it met the intentions. Students received feedback during the project through the continuous formative assessment described above, allowing them to improve their work. Summative assessment of the quality of the final product was completed by applying a rubric that I have used in this class since 2003. The rubric (Appendix I of Kelley and Visaggi, 2012) allocates 50 points among the title (2 points), abstract (4), introduction (6), methods (4), results (4), discussion and conclusions (18), and format (12); introduction, methods, and results (14 points total) were group authored, with students responsible individually for all other components graded by the rubric. I also annotated each manuscript with comments, so students could ascertain the areas in which they had done well and those in which they had fallen short of expectations. The final reflection pieces – by the students and me – represented an excellent approach to evaluation of the overall experience (see below). See also Kortz and Kraft (2016), for comments on the usefulness of reflection in an undergraduate course-embedded research project.

3.1.8 Acknowledgment

The final principle articulated by NSEE (1998) is that progress and accomplishment should be recognized throughout and at the end of the experience to provide closure. In IP, acknowledgment occurred through one of the key aspects of undergraduate research experiences recognized by the National Academies (2017): communication of results. The last day of class, students presented their results to one another and we drew conclusions together relating to the scientific outcomes of the study. This venue allowed me to acknowledge the work they had done in addressing the scientific issues presented in the initial handout. The ultimate recognition was coauthorship of a published abstract, which students could include in their resumes. Students who attended the national GSA meeting to present this work received further acknowledgment from the paleontological community (external validation, per Corwin et al., 2015). Similarly, in a service-learning inquiry project that examined water quality of a local lake, Koretsky et al. (2012) reported that students viewed as a reward (acknowledgment) the opportunity to present their results to the local community.

3.2 Was It Worth It? (AKA Assessment)

Research projects embedded in courses are a lot of work, for faculty and students. Students in IP required extensive mentoring and monitoring. The idea of doing authentic research within the context of a course was entirely new to most IP students; most had not conducted "real" research before, and some had never written a scientific paper. One student commented in the final reflection paper:

> I had never done a project like this before . . .. This project was a fantastic experience for me and it really showed me what it takes in order to success-fully complete a project like this . . .. The part of the experience that I had the most trouble on was drawing conclusions from the data and discussing the results at the end due to my lack of experience in this field, but I worked hard and I believe that in the end I understood what to do.

Another challenge for the students was working in teams, because no student's individual report could be completed without incorporating results of data collection and analysis by other students. Each student had to submit results promptly or incur the ire of teammates (indeed, evaluations by peers can be a motivating factor in terms of group work; see, e.g., Tessier, 2012). Thus, I needed to monitor student progress to mitigate any team dysfunction – a problem I encountered the first time I used this approach in IP (Kelley and Visaggi, 2012). In their final reflection papers, some students commented on the challenges of group work. One noted, "It was difficult working in a group when it came to the paper. It was hard to get the information that was needed on time to complete the paper." Another expressed frustration about teammates' writing contributions:

> As a whole, our group did exceptional work when it came to combing through the thousands of specimens and identifying taxa. However ... other group members seemed to only put in the bare minimum concerning the quality of their sections of the paper. [Another student] and I spent hours correcting mistakes and poor use of the English language.

Meeting various deadlines for the completion of tasks also was a challenge for the students, especially because they were enrolled in other geology courses with overnight field trips late in the semester. One student wrote:

> [T]he process of separating, cleaning, identifying, and counting all the specimens took about a month longer than anticipated. The data gave us some skewed numbers which we had to decipher and interpret before we could even begin writing about it. This gave us a relatively short window to write our findings into a formal research paper. But we got it done and we got it done well.

3.2.1 Approaches to Assessment

Because embedding a research project, especially one of such magnitude, within a course is an ambitious undertaking for students and faculty, assessment of the utility of this approach is important. I found the student reflection papers to be most helpful in such assessment. (It should be noted that the reflections did not contribute to student grades but were sent to the ETEAL assessment office. Thus, the student comments are likely to be a true indication of their experience in the course and were not written to please their instructor.) In addition to the reflection pieces, the continued engagement of the students in paleontology is an indicator of the utility of the approach (see also Corwin et al., 2015). Data on student performance on the project, as judged by the grading rubric, also reveal whether learning objectives for the project were met. Both retention and progress determined from student products were considered appropriate measures of success by National Academies (2017).

3.2.2 Conclusions from Student Reflections

Despite the challenges noted above, students seemed to appreciate the opportunity to participate in actual scientific research, using samples that no one had ever seen, let alone analyzed, before. The student reflections indicated that, in general, the initial goals and intentions were met, both in terms of academic enhancement and personal growth. The students were very excited about becoming coauthors of a published GSA abstract and presentation, as several stated in their final reflection papers.

Student reflections demonstrated the value of authenticity. Students appreciated being able to do "real" research that would matter to others beyond their classroom and beyond the university (external validation; Corwin et al., 2015):

> When I first began this course we were informed that we would be assembling a research paper with the intention of being published. This was the first time I would be writing a paper of this caliber and knowing that it was going to be published and presented at GSA had me more than just a little excited . . .. I think it's important for students to be able to experience a class and situation such as the one ours experienced. It helps us to be prepared for the world that lies ahead of us after our schooling is complete. I believe we are better-rounded as scientists and thinkers and that will give us give us an advantage as we pursue our careers of choice.
>
> Although at times I thought to myself, "What am I *doing*, counting thousands of one kind of shell? This is insane," I realize now the significance of tedious work: noteworthy and intriguing results and experience. I was able to work with actual fossils and make interpretations about a million-year-old marine community, something I might never be able to do again.
>
> This class has helped me grow as a student and person . . .. It has made it easier for me to work as a geologist by giv[ing] me a research project that applies to paleontology in a realistic setting.

Students also stressed the utility of the research project in teaching them about the scientific process, instilling research skills, and reinforcing class content by applying paleontological concepts. Their experience thus aligns well with the characteristics of undergraduate research experiences identified by the National Academies (2017).

> This class reinforced the use of the scientific method, scientific literature research, field collection techniques, data management and interpretation, and concepts learned in physical and historical geology. This was especially evident in our research project concerning drilling predation in predatory gastropods. In addition to our field trip to New Bern, North Carolina, this was my favorite experience as a learning geologist.
>
> The class project we performed provided me with a tremendous amount of knowledge about the general scientific process of research and how to effectively budget my time appropriately . . .. Maybe I will not be studying fossilized invertebrates, however, the scientific process of research and writing is one that will benefit me no matter where I end up.
>
> Use of the scientific method and scientific writing skills was extremely important for this project, especially for the portion in which students had to

draw conclusions about the fossil community based on their data ... This research project has been and will be useful to me as it was instrumental in continuing to fill out my research skill set.

> For this project it was very important to be meticulous about the details. From the very first day of sieving to the last count of *Mulinia*, it was important to do things right. In this way, we learned a lot about the scientific method and its applications to our project. I feel that we, as a class, now have a better understanding of the "real feel" of research .... I believe one of the most importan[t] aspects of the assignment was being able to extrapolate information out of all the data that was collected. To me, this was the most fundamental and exciting component. All semester long we learned about key concept[s] of invertebrate paleontology and were then able to apply what we learned.

Students also commented on the value of working in teams (see also National Academies, 2017). Despite the challenges of group work, students also recognized its benefits:

> It was very important to work together and communicate to one another. In this way it made it easier to accomplish tasks. Sorting the sample by specimens was, by far, the most daunting task that we had, but by assigning "jobs" every period we were able to see ourselves through. I believe that my group did a great job in the tasks that we were assigned. Everyone was ready and willing to help each other.

> I felt like working in a group was good. It was helpful when working with the large sample. It helped me work on my social skills in a work environment .... My skills of being easy-going and helpful ... gave me the ability to work with all different kinds of people and be successful. I feel our group was very productive and helpful to one another.

Personal satisfaction and a sense of accomplishment also derived from the experience; one student commented, "My research project is one of my greatest achievements being that I've never written a paper so long." Another summed up this attitude eloquently:

> I have learned much about perseverance, thoroughness, and accuracy. The process of sorting and counting the sample was tedious work. But, after all the data was collected and processed, we were rewarded with a quality study and a quality paper we could all be proud of. Scientific work is not always as straight forward as you expect it to be .... This is something we experienced firsthand and I think it is an invaluable lesson to learn. At some other point(s) in our careers we will assuredly come up against the same sort of challenge. It feels good to already have this experience under our belts and I thank Dr. Kelley for giving us this great opportunity .... This course was hard work, but it was fun. If given the choice, I'd still have taken it instead of something else; furthermore, if

given the opportunity, I'd like to take a higher level course that might stem from it.

3.2.3 Retention

Another way to gauge the success of this pedagogy is by the retention of students in paleontology after the completion of the course (Kelley and Visaggi, 2012; see also Corwin et al., 2015, and National Academies, 2017, for use of persistence in STEM as a measure of success of undergraduate research experiences). The student quoted above indicated a desire to take more paleontology courses – and he did (my stratigraphic paleontology course). This student was not the only one who remained engaged in paleontology based on the experience in IP. Over half the students (50 of 99) enrolled in IP from 2003 to 2015 remained involved in paleontology following the course by presenting the work at a GSA meeting and/or a research showcase on campus (38 students), or continuing their research as a directed individual study, taking additional paleontology courses, or pursuing graduate study in paleontology (22 students; note that some students who were presenters also pursued further study in paleontology). In fact, three of my students who participated in my in-course research projects now have PhD's and have used a similar approach in their courses (see for example Visaggi et al., 2014). Unfortunately, no comparable retention data exist for the IP course prior to introduction of this pedagogy, because before 2003 the course was taught by other faculty.

3.2.4 Student Performance

An additional measure of the effectiveness of the learning experience is the performance of students on the research project (Kelley and Visaggi, 2012) based on scores using the grading rubric. Because the same detailed rubric was used each year in grading the papers resulting from the project, the grading protocol was consistent from year to year, allowing comparisons among cohorts (Leydens and Santi, 2006; Kelley and Visaggi, 2012). Students who performed well on other aspects of the course (three exams and the "traditional" laboratory exercises, each accounting for 20 percent of a student's grade) also achieved high scores on the project. Spearman's rank correlation coefficient between students' project grades and their grades on all other work was 0.4776 (97 d.f., $p \ll 0.001$). However, the average class grade on the project exceeded the average class grade on all other work for every year that the course was taught, with the exception of the first two cohorts. A Mann-Whitney U test indicated that the difference between project grades and grades on all other work was statistically significant ($Z = 1.77$, 7 d.f., $p = 0.0384$). These results suggest that students who did not do as well in a testing situation were nevertheless able to succeed in the research project. Alternatively, it could be argued that students whose performance on class work was weaker might have had project grades inflated by the contributions of group members. However, the rubric was heavily weighted (72 percent) towards individually written components, and each student's contributions to the group sections were identifiable based on the tasks assigned within the team (e.g., diversity, life mode, or predation analysis). Thus the project grades are useful indicators of each student's performance.

Project grades were significantly lower ($t = -3.13$, 97 d.f., $p = 0.001$) for students in the first two cohorts (2003, 2005) than in subsequent years that the course was taught (2006–2015), although grades in the rest of the course did not differ ($t = 0.01$, 97 d.f., $p = 0.495$). These results suggest that the research project was a less effective pedagogy when initially employed. Lessons learned concerning orientation and training of students and monitoring of team function enabled me to correct issues in project implementation that I noted in the first two years I used this pedagogy.

3.3 Challenges and Benefits for Faculty

Challenges of implementing course-based research for undergraduates are manifold. Undergraduates are typically inexperienced in research, necessitating training and monitoring throughout the project; such monitoring may indicate that changes in project goals and schedule are warranted, requiring faculty to be flexible. In addition, student work needs to be checked for accuracy. In 2013, for instance, because of the unanticipated vast number of specimens in the samples, I had to put in extensive time after the semester ended to vet the results and guarantee that they would be publishable. Group work by students also presents its own set of challenges (see also Burke, 2011). Teams need to be monitored as the research is conducted, which is facilitated by the instructor working alongside the students each lab period. Some years, teams differed in the pace at which work on their samples proceeded, and I asked members of teams making faster progress to help other teams, so that the class could move forward together. In addition, grading group projects has challenges if some students contribute more than others. Working in the lab alongside the students encourages greater effort and prevents individual students from shirking their responsibilities. In addition, the grading problem is mitigated because individually written sections form the major component of the final papers.

Considering all these challenges, is this approach worth it to the faculty member? Other than the satisfaction of improving the student learning experience (National Academies, 2017), what gains does course-embedded research provide for the instructor?

Corwin et al. (2015) noted that one attraction of teaching CUREs is the connection that faculty can make to their research interests. I have found the approach useful in addressing questions in my own research that require massive data collection. Research that requires processing samples from multiple localities and/or stratigraphic levels is particularly amenable to the iterative nature of the IP class projects – though publishing with students is more work than doing it on your own. Nevertheless, the nine IP classes I taught from 2003 to 2015 resulted in 18 abstracts and presentations and several peer-reviewed research articles (Kelley and Hansen, 2007; Kelley, 2008) and pedagogical publications (Kelley, 2004; Kelley and Visaggi, 2012 – and the present article!). Indeed, faculty who carry heavy teaching loads may find this approach the most effective way to get research done.

In general, benefits for faculty will vary depending on the departmental and institutional context, including their culture and mission. For instance, some institutions expect faculty to involve undergraduates in research, whereas at other institutions such time-intensive activities may detract from other activities (e.g., grantsmanship) that are considered more valuable in the tenure and reward system (National Academies, 2017). At UNCW, this approach coincided well with the university's applied learning emphasis. Faculty rewards are likely to be greatest at institutions with an established culture of undergraduate research.

4 What If You Don't Have a Whole Semester – Or 23,276 *Mulinia*?

Not all courses are amenable to implementing the type of research experience described here. For instance, the intensive mentoring required makes offering such experience in higher-enrollment courses challenging. As discussed by Kelley and Visaggi (2012), involvement of additional appropriately trained mentors (teaching assistants or advanced students who wish to gain mentoring experience) would permit offering course-embedded research experiences in classes with higher enrollments. This approach has been used successfully by Visaggi (see, e.g., Visaggi et al. 2014) in classes enrolling more than 20 IP students.

In addition, time constraints may prohibit including a semester-long course-based research project. Class time spent on research reduces that available for "presentation of basic concepts and principles" (Foos, 1997, p. 322). However, even if time constraints prevent spending an entire semester on a class research project, research experiences can be incorporated within a course to a more limited degree. I provide examples of this approach for two courses I have taught.

In historical geology each year, we took a field trip to a local Pleistocene marl quarry, where I instructed students to collect as great a diversity of species as they could find. During the next few laboratory sessions, students worked together and with me to identify and collect data from their fossils. I distributed each student's faunal list (after vetting it) to the rest of the class, and they were responsible for determining each species' life mode (e.g., diet, mobility) from online paleontological databases (Neogene Marine Biota of Tropical America, 2016; Paleobiology Database, 2017). They also studied their specimens for evidence of organismal interactions (holes drilled in shells by predatory gastropods, growth of corals on mollusc shells, etc.). Each student wrote a report, with graphs of diversity and life mode, interpreting the paleoenvironment and paleoecological interactions based on the fauna. Although the data were not rigorous quantitatively, students were introduced to the techniques and processes involved in authentic research. For another example, see Science Education Resource Center (2016), which documents a similar field-trip-based project employed in Invertebrate Paleontology by Mark Wilson at the College of Wooster.

I also embedded a short-term paleoecology research project in an Honors course I team taught with a colleague in psychology, "Behavior of Animals: Dead and Alive." The course focused on methods used to quantify or describe behavior of living animals and to infer it in dead (i.e., fossil) animals. Students completed a research project on behavior of live animals they could observe outside of class in the college environment (squirrels, birds, fish in a tank). The fossil research project involved a field trip to a local quarry. Groups of students worked together in the field to collect fossils and develop hypotheses about the behaviors of the animals represented. During the following two weeks of class time they tested their hypotheses by collecting data on molluscan morphology and trace fossil occurrence. Each group presented its results in a PowerPoint presentation to the class the final days of the semester (see Honors 120 field trip assignment, Appendix 2).

Even such short-term research projects are valuable, as indicated by student comments in reflection papers. Students in the animal behavior course found the hands-on, minds-on research approach to be a useful learning experience because they could apply information they learned in class. One student commented, "The great thing about this class was that after we learned all the methods of identifying behaviors, we got to put them to use." Another noted, "Before this class, I had not really considered that a hypothesis could be tested using dead animals, but the field trip to the quarry and subsequent project allowed us to see how this could be done, and try it ourselves." Others stated:

> Taking this Honors 120 class was a great idea for my first semester in college. I really enjoyed learning about animal behavior in an experimental way, rather than in a lecture hall. With this style of learning I was able to figure out an animal's behavior first hand and receive true field experience both with dead animals and live animals.

> While we learned a lot about different methods of observing and recording animal behavior in the classroom over the course of the semester, I definitely feel like I learned the most while in the field . . .. I believe the quarry was the best learning experience for these methods. Overall I learned a lot about observing, recording, and deciphering animal behavior from this course. It provided me with hands on experiences I may not have been able to participate in as easily outside of the class, and I'm glad I picked this seminar.

Students also were more engaged in the topic as a result of incorporating research activities, which they characterized as "fun." For instance, a student commented, "The dead animal project was a lot of fun because we were able to gather a lot of data on a topic and see how they behaved, or in my case, to see if they showed preference of shell shape." In the same vein, one noted:

> The most fun part of the class was visiting the quarry . . .. [We] were able to analyze all the shells, determining what the species was and coming up with a hypothesis. After [that] we were able to test our hypothesis by conducting an experiment and looking at the data we had collected. I truly learned so much from this part of the class.

Students also were more invested in the research projects because they were given the opportunity to develop their own hypotheses and design the methods for testing them (see also Mayborn and Lesher, 2000, and National Academies, 2017):

> The final dead and live animal projects involving the fossil dig and the experiments with Betta Fish also provided good hands on experience observing animal behavior using the methods that we learned in class. I really liked these projects because they allowed us to study an aspect of animal behavior which we chose based on our own personal interests. I really liked this freedom to test my own hypothesis and formulate my own methods of experimentation.

These short-term class research projects benefited from being able to incorporate field work, but not all institutions are in proximity to appropriate field sites (Kortz and Kraft, 2016). Successful specimen-based research experiences, both long- and short-term, can be designed using pre-existing research collections. For instance, Montgomery and Donaldson (2014) described a problem-based honors introductory course in paleontology that made use of Montgomery's research collections from Big Bend, Texas. Instructors without an active collections-based research program who wish to employ specimen-based research might seek collaborations with other researchers or museums. For example, based on my involvement of classes in research, I was invited to use bulk samples from the American Museum of Natural History for class projects (we didn't have 23,276 *Mulinia*, but several graduate and undergraduate classes and students doing directed individual studies did count 16,685 specimens of another small bivalve, *Transenella*!). However, faculty who do not relish checking identifications for thousands of bivalves (or brachiopods, or forams) may prefer to design authentic research projects that employ online datasets or databases. Instructors with limited resources, in particular, may wish to use such minds-on projects, which are the focus of Elements in this series by Cohen et al. (2018) and Lockwood et al. (2018).

5 Conclusions

A wide spectrum of approaches exists for incorporating research into undergraduate teaching, including research-led teaching that involves transmission of current research findings and research-oriented teaching, in which students learn research techniques. The greatest benefits to students accrue from research-based teaching, in which students conduct research. Students thereby acquire an understanding of the research process; they develop research and technical skills, improve their abilities to think critically and solve problems, learn to work in teams and independently, and become more self-confident. Research projects, both short- and long-term, embedded in courses extend these benefits to a broader population than do apprentice-style research experiences. Such course-embedded research is most effective when characterized by intentionality, planning, authenticity, reflection, training, monitoring, assessment, and acknowledgment.

Appendix 1

Instructions for GLY 337 Invertebrate Paleontology "ETEAL" Research Projects, Spring 2013

GLY 337 is an "early adopter" course for UNCW's new "ETEAL" program – Experiencing Transformative Education through Applied Learning. In applied learning, students "integrate theories, ideas, and skills that they've learned in new contexts, thereby extending their learning" (www.uncw.edu/qep/overview .html). Although GLY 337 has involved applied learning for the past decade, this semester we are serving as an example to the campus of how applied learning, specifically authentic research in the context of a course, can enhance students' education. As part of our project, you will be asked to reflect on your learning experience.

Scientific Context

Our research ties in to the work that has been conducted in UNCW's "Research Experiences for Undergraduates in Biodiversity Conservation" program (summers 2008–2010). My classes at UNCW (GLY 337, 510, 533) have also been involved in this research over the past few years. The REU goal has been to examine changes in the marine ecosystem of the Carolinas over the past 3 million years, in order to compare ongoing human-induced changes with the changes produced by natural disruptions to the system during the Plio-Pleistocene. By understanding the natural variability of the system and the responses of the organisms (to the Plio-Pleistocene extinctions) we can better understand the effects of humans on the environment and the health of the modern ecosystem. This program is part of a brand-new field known as conservation paleobiology.

What do we know about the Plio-Pleistocene extinction? Information from other regions (Florida, Virginia, and the Caribbean) has suggested that:

- it affected approximately 60 to 65 percent of the molluscan fauna in the western Atlantic
- it likely occurred in several stages
- causes are controversial but may include changes in temperature, productivity, and sea level

- there is some evidence for productivity change in the Caribbean based on changes in community structure
- in Florida, diversity recovered because extinctions were balanced by originations

Comparable work had not been done in the Carolinas, but based on species lists it was suggested that species diversity did not recover in the Carolinas following the extinction.

The REU program and UNCW classes have analyzed bulk samples from 17 sites. Localities from the Neuse River northward include an unnamed Pliocene unit (Yorktown Formation?) at Fountain, North Carolina, the Chowan River Formation at its type locality, the lower James City Formation at Lee Creek, and upper James City Formation near its type locality. In southeast North Carolina and adjacent South Carolina we sampled four Duplin Formation, five lower Waccamaw Formation, and three upper Waccamaw Formation sites. Last semester the GLY 533 class added material from the Canepatch Formation in South Carolina (same locality as you will be studying, but different part of the section).

The stratigraphic relationships among the units we have been investigating are shown below (diagram provided by Dr. William Harris).

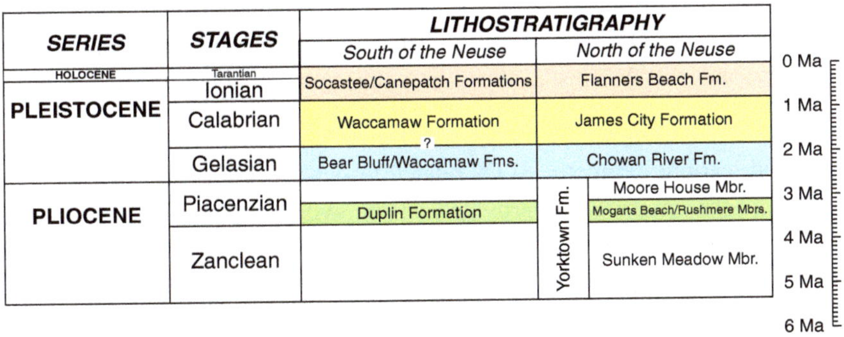
SERIES
STAGES
LITHOSTRATIGRAPHY
South of the Neuse
North of the Neuse
HOLOCENE
Tarantian
Ionian
PLEISTOCENE
Calabrian
Gelasian
PLIOCENE
Piacenzian
Zanclean
Socastee/Canepatch Formations
Flanners Beach Fm.
Waccamaw Formation
James City Formation
?
Bear Bluff/Waccamaw Fms.
Chowan River Fm.
Moore House Mbr.
Mogarts Beach/Rushmere Mbrs.
Duplin Formation
Yorktown Fm.
Sunken Meadow Mbr.
0 Ma
1 Ma
2 Ma
3 Ma
4 Ma
5 Ma
6 Ma

Our results have been surprising.

- Multiple phases of extinction don't seem to be supported by genus-level analyses.
- Diversity shows relatively little change until the upper Waccamaw/James City formations.
- Community structure shows relatively little change.

These results are tentative, because more data are needed, especially from the upper Waccamaw Formation, thought to have been deposited after the extinction.

This semester's work is particularly important because we are adding a new locality of the Waccamaw for which the molluscan fauna has not been studied in depth before. The locality is the Windy Hill airstrip (Dubar's WA 56) locality, Myrtle Beach, Horry County, South Carolina. We are not even sure if the samples represent the lower Waccamaw or the upper Waccamaw. The work done by this class may help determine the stratigraphic position of the unit.

Additional questions this class can help address:

- What is the history of diversity change in the Carolinas? Was there a drop in diversity between the lower and upper Waccamaw formations?
- How did community structure change over time, particularly trophic/guild structure?
- Is there a "decoupling" between diversity and other ecological parameters?
- How did particular ecological interactions (such as drilling predation) vary through time?
- How do changes through time compare to spatial variation in the system?

Student Learning Outcomes and Critical Reflection

This project will contribute towards fulfilling the Student Learning Outcomes (SLOs) for an Information Literacy and Writing Intensive course (see syllabus for IL and WI SLOs). You will receive instruction from Peter Fritzler (Randall Library) and Will Wilkinson (Writing Center) that will help you achieve the IL and WI SLOs.

Applied learning promotes SLOs related to our specific course content also. Consider the course SLOs provided on the syllabus. How does our project relate to each of the course SLOs?

Finally, in addition to increasing content knowledge, applied learning projects such as this one promote a variety of life skills, including improved critical thinking and problem-solving skills and enhanced oral and written communication skills. In addition, gains in personal development occur, as students are better able to work independently as well as collaboratively and also increase self-confidence.

The following ETEAL SLOs are shared by all students participating in ETEAL experiences at UNCW (www.uncw.edu/qep/assessment.html):

1. Articulate their expectations, the purpose, and/or the goals of the experience in terms of their personal educational development. [Thoughtful Expression]

2. Synthesize knowledge drawn from their coursework to address the issues/challenges/questions involved in the experience. [Critical Thinking, Inquiry]
3. Communicate the impact or significance on their personal educational development and on others in the profession or in the field at the conclusion of the experience. [Critical Thinking]

You will achieve ETEAL SLO 1 at the onset of the applied learning experience as the laboratory assignment on 1/16/13.

ETEAL SLO 2 will be achieved by completing the research project and report (see below).

ETEAL SLO 3 will be achieved by completing the final reflection paper (see below).

Instructions for Completing the Research Project

In order to answer questions about the Plio-Pleistocene extinctions, we will need to collect data comparable to those collected by the REU and UNCW students from 2008 to 2010 and compare our data to results from other Plio-Pleistocene stratigraphic units and localities.

During the course of the semester you will complete the following steps:

1. Remove and save all whole specimens, bivalves that contain beaks, and gastropods that contain apices. (If you find a piece of a gastropod that does not include an apex but appears to be a unique taxon, retain that also – ditto for bivalves.) Broken specimens that are not unique and that lack beak/apex should be retained in a bag labeled "junk."
2. Separate all whole specimens, beaks, and apices into different genera.
3. Identify each genus using the references provided; you may also use previously studied samples as a starting place. Make a tag for each specimen (out of scratch paper) stating specimen ID and the reference/page on which you found it so you can recheck if needed. After Sam and I have checked your IDs you can fill out a more permanent label (I will provide the labels).
4. Develop hypotheses for testing using these samples (we will work on this as a class).
5. Data collection
   - All team members will work together to count:
     o Number of whole or nearly whole bivalve valves and gastropod shells
     o Number of fragments that contain the beak (bivalves) and apex (gastropods)

Appendix 1

- o Number of whole specimens with complete drillhole
- o Number of whole specimens with incomplete drillhole
- Counts will be written IN PENCIL on the tag in the box for each genus.
- Write the data on a datasheet as taxa are counted (I will supply data sheets) and then enter the handwritten data into an Excel spreadsheet and email it to me for distribution.

6. Data analysis
  - Drilling predation data (use only the whole or nearly whole specimens) and calculate for each genus and for the bivalve and gastropod assemblages as a whole
    - o Drilling frequency (DF) for bivalves = # valves with complete drillhole/half the # valves
    - o Drilling frequency (DF) for gastropods = # shells with complete drillhole/total # shells
    - o Prey effectiveness (PE) = # Incomplete Holes/# Total Attempts (complete and incomplete)
    - o Present these data in a table and graphically
  - Life mode analysis
    - o Compile data for each genus based on the NMITA database at http://nmita.geology.uiowa.edu/database/mollusc/mollusclifestyles.htm, including
      - ▪ Feeding information
      - ▪ Relation to substrate (bivalves only)
      - ▪ Mobility (bivalves only)
      - ▪ Attachment (bivalves only)
    - o Calculate the percent of genera represented by each life mode and produce graphs (histograms or pie charts) and tables to portray your results
  - Diversity information
    - o Richness (# of genera)
    - o Use rarefaction to compare richness between samples www.uga.edu/~strata/software/Software.html
    - o Produce rarefaction curves and determine 95 percent confidence intervals
    - o Compare your results with those from previous studies of different localities (I will provide these data).

You may divide the work in whatever way seems feasible. For instance, one or more members may specialize on drilling predation, other(s) on life mode, other(s) on background research and stratigraphy, etc. We can make these decisions after we see how diverse your samples are and how much time will be required for each phase of study. I will meet with the individuals from each team who are assigned to a given task to get you started on your assignment.

7. Produce a final written report of your findings following the format of a scientific paper (see attached sheet). Each of you should complete and hand in your own paper, but team members should cooperate with one another in writing the paper. Each section should have authorship as follows:
   - Title – each person should write his/her own title.
   - Abstract – each person should write his/her own abstract.
   - Introduction – the background/stratigraphy researcher should write this section.
   - Materials and methods – the background/stratigraphy researcher should write the "materials" section for each team, and each team member should contribute a description of the methods for his/her own part of the analysis.
   - Results – this section should have different subsections, written by the persons assigned to drilling frequency, life mode, and diversity analyses. Provide any figures relevant to your analyses.
   - Discussion and conclusions – each person should write his/her own discussion and conclusions based on the combined results.
   - References cited – if you cite any references in the section you have written, please provide those citations to the other members of your group. Each person will put together his/her own references cited section.

8. Final reflection paper can be handed in with the written report or at the final exam.

Your final reflection paper should consist of two sections that consider the following:

1) Academic enhancement
   - What relevant theories, ideas, and skills were you able to apply during this experience or that helped guide this experience?
   - How did this experience help you understand concepts in paleontology?
   - What academic skills did you use during this experience (e.g., ability to use the scientific method, read and understand scientific literature, collect and analyze data and interpret the results, scientific writing skills)?

Your reflection should emphasize what you learned and how the applied learning experience helped you learn it. Some prompts for yourself are:

"I learned that ..., " "I learned this when ..., " "This learning matters because ..., " and "In light of this learning I will ... "

2) Personal growth

- What expectations did you bring to this experience? To what extent were they met?
- How do you feel your group performed?
- How do you feel you performed within your group? In what ways did you do well and what personal characteristics helped you be successful?
- What aspects of this experience were difficult, and what contributed to these difficulties?
- Analyze the implications of what you learned from this experience for your future.

Project timeline:

Tues 2/27   IDs completed and counts started

Tues 3/13   Counts completed and data recorded on datasheet

Mon 3/25   Data entered into spreadsheet and emailed to PK

Tues 4/3   Analyses completed

Thurs 4/11   Written sections (with graphs) submitted to other team members and PK for critiquing

Thurs 4/25   In-class discussion of results

Thurs 4/25   Papers due at 5:00 pm; late papers will be penalized!

Tues 4/30   Final reflection due (Final Exam day)

A few references to get you started (Peter Fritzler will help you find others on the local stratigraphy and depositional history, the extinction episode and its effect on diversity and ecological structure, drilling predation, etc.):

Allmon, W. D., G. Rosenberg, R. W. Portell, and K. S. Schindler. 1993. Diversity of Atlantic coastal plain mollusks since the Pliocene. *Science* 260:1626–1628.

Du Bar, J. R., and Furbunch, H. W. C. 1965. The Waccamaw Formation (Pliocene?) and its Macrofauna, Intracoastal Waterway, Horry County, South Carolina: *South Carolina Devel. Board Div. Geology Geol. Notes* 9(1): 1–24.

Kelley, P. H., and T. A. Hansen. 2003. The fossil record of drilling predation on bivalves and gastropods, pp. 113–139. In P. H. Kelley, M. Kowalewski, and T. A. Hansen (eds.), *Predator-Prey Interactions in the Fossil Record*. Kluwer Academic/Plenum Press.

Kowalewski, M. 2002. The fossil record of predation: an overview of analytical methods, pp. 3–40. In M. Kowalewski and P. H. Kelley (eds.). *The Fossil Record of Predation*. The Paleontological Society Papers, vol. 8.

Todd, J. A., J. B. C. Jackson, K. G. Johnson, H. M. Fortunato, A. Heitz, M. Alvarez, and P. Jung. 2002. The ecology of extinction: molluscan feeding and faunal turnover in the Caribbean Neogene. *Proc. R. Soc. Lond. B* 269:571–577.

Ward, L. W., R. H. Bailey, and J. G. Carter. 1991. Pliocene and early Pleistocene stratigraphy, depositional history, and molluscan paleobiogeography of the Coastal Plain. In J. W. Horton, Jr. and V. A. Zullo (eds.), *The Geology of the Carolinas: Carolina Geological Society fiftieth anniversary volume*. Knoxville, Univ. Tennessee Press.

Appendix 2

HON 120 Behavior of Animals Field Trip Assignment

On Saturday, October 20, 2012, you will visit a quarry near Old Dock, North Carolina. The quarry exposes sediments of the Waccamaw Formation, deposited approximately two million years ago during the Pleistocene Epoch when sea level was higher than today. The Waccamaw sands are rich in molluscs, primarily bivalves (clams, oysters, scallops) and gastropods, including predatory moonsnails (shell drillers) and whelks (which chip or wedge open shells of clams, sometimes injuring their own shell in the process). You may also find sand dollars, crab claws, corals, bryozoans, and shark teeth.

The goal of this project is to observe evidence of invertebrate behavior and draw and test hypotheses about behavior based on the fossil record. Your guides will point out examples. You may observe evidence of behaviors such as:

- Behavior related to reproduction and recruitment, e.g., stacking of *Crepidula fornicata*, settling of larvae (growth of corals, bryozoans, or other animals on shells)
- Occupation of shells by hermit crabs (crabs don't preserve but you might find evidence of shells that were "hermited")
- Competition by encrusting organisms overgrowing other encrusters on shells
- Evidence of attacks by crabs on molluscs, especially if the victims survived and repaired their shells
- Evidence of feeding behavior by moonsnails (drillholes!)
- Evidence that whelks broke their shells during feeding, followed by repair

Your assignment:

1) How many types of behavior can you find evidence for? Collect representatives of as many types of behavior as possible. We will discuss these in class later this semester.

2) Group project: each group should develop and test a hypothesis related to one of the above behaviors. Shell drilling is probably the behavior most amenable to studying, though you are welcome to study other behaviors. Possible questions to investigate:

   a. Do particular types of larvae show a preference for settling on particular types of organisms?

b. Do moonsnails prefer to drill certain species of prey? What characteristics might make a prey item desirable or undesirable?

c. Do moonsnails prefer to drill certain areas of the prey shell?

d. Do moonsnails prefer to drill certain sizes of prey?

e. Do large whelks show more repair scars than small whelks?

Based on one of these questions (or another one – get innovative if you like!),
develop a hypothesis for testing, e.g., moonsnails prefer to eat small prey
(why?). Collect the specimens you will need to answer your question (you
might want to focus on collecting a particular prey species). You will take your
specimens back to UNCW with you for data collection (e.g., measurement of
drillhole size, prey size, proportion of shells that are drilled). Make sure you
collect enough specimens to answer your question!

References

Allen, J. L., Creamer, E. G. and Kuehn, S. C. (2017). Testing the impact of a multi-year, curriculum-based undergraduate research experience (MY-CURE): observations from a 5-semester cohort. *Geological Society of America Abstracts with Programs*, **49**(6), doi: 10.1130/abs/2017AM-301844.

American Library Association. (2015). Framework for information literacy for higher education, www.ala.org/acrl/standards/ilframework, accessed 15 October 2017.

Appedoe, X. S., Walker, S. E. and Reeves, T. C. (2006). Integrating inquiry-based learning into undergraduate geology. *Journal of Geoscience Education*, **54**, 414–521.

Auchincloss, L. C., Laursen, S. L., Branchaw, J. L., Eagan, K., Graham, M., Hanauer, D. I., Lawrie, G., McLinn, C. M., Pelaez, N., Rowland, S., Towns, M., Trautmann, N. M., Varma-Nelson, P., Weston, T. J. and Dolan, E. L. (2014). Assessment of course-based undergraduate research experiences: a meeting report. *CBE Life Sciences Education*, **13**(1), 29–40.

Bloom, B., Englehart, M., Furst, E., Hill, W. and Krathwohl, D. (1956). *Taxonomy of Educational Objectives: The Classification of Educational Goals. Handbook I: Cognitive Domain.* New York: Longmans, Green.

Brew, A. (2010). Imperatives and challenges in integrating teaching and research. *Higher Education Research and Development*, **29**(2), 139–150.

Budd, D. A., Kraft, K. J., McConnell, D. A. and Vislova, T. (2013). Characterizing teaching in introductory geology courses: measuring classroom practices. *Journal of Geoscience Education*, **61**, 461–475.

Burke, A. (2011). Group work: how to use groups effectively. *The Journal of Effective Teaching*, **11**(2), 87–95.

Carleton University. (2017). Tips for Incorporating Research into Teaching, carleton.ca/edc/wp-content/uploads/Incorporating-Research-into-Teaching.pdf, accessed 15 October 2017.

Cohen, P. A., Lockwood, R. and Peters, S. (2018). Integrating Macrostrat and Rockd into undergraduate earth science teaching. In P. Cohen, L. Park Boush, and R. Lockwood, eds., *Pedagogy and Technology in the Modern Paleontology Classroom. Elements of Paleobiology*, 1, **-**.

Corwin, L. A., Graham, M. J. and Dolan, E. L. (2015). Modeling course-based undergraduate research experiences: an agenda for future research and evaluation. *CBE Life Sciences Education*, **14**(1), es1. http://doi.org/10 .1187/cbe.14–10-0167.

Darden, A. (2003). Integrating research and teaching heightens value to and of undergraduates. *American Society for Microbiology News* **69**(7), 331–335.

Davies-Vollum, K. S. (2006). Using grain size analysis as the basis for a research project in an undergraduate sedimentology course. *Journal of Geoscience Education*, **54**, 10–17.

Deslauriers, L., Schelew, E. and Wieman, C. (2011). Improved learning in a large-enrollment physics class. *Science* **332**, 862–864.

Ellwein, A. L., Hartley, L. M., Donovan, S. and Billick, I. (2014). Using rich context and data exploration to improve engagement with climate data and data literacy: bringing a field station into the college classroom. *Journal of Geoscience Education*, **62**, 578–586.

ETEAL. (2018). About ETEAL, https://uncw.edu/eteal/about.html, accessed 6 June 2018.

Foos, A. M. (1997). Integration of a class research project into a traditional geochemistry lab course. *Journal of Geoscience Education*, **45**, 322–325.

Gold, A. U., Kirk, K., Morrison, D., Lynds, S., Sullivan, S. B., Grachev, A. and Persson, O. (2015). Arctic Climate Connections Curriculum: a model for bringing authentic data into the classroom. *Journal of Geoscience Education*, **63**, 185–197.

Gonzales, D. and Semken, S. (2006). Integrating undergraduate education and scientific discovery through field research in igneous petrology. *Journal of Geoscience Education*, **54**, 133–142.

Griffiths, R. (2004). Knowledge production and the research–teaching nexus: the case of the built environment disciplines. *Studies in Higher Education*, **29**(6), 709–726.

Gutiérrez, M. and Baker, B. (2013). Making connections to real data and peer-review literature: a short soil exercise in a geochemistry class. *Journal of Geoscience Education*, **61**, 53–58.

Hackathorn, J., Solomon, E. D., Blankmeyer, K. L., Tennial, R. E. and Garczynski, A. M. (2011). Learning by doing: an empirical study of active teaching techniques. *The Journal of Effective Teaching*, **11**(2), 40–54.

Healey, M. and Jenkins, A. (2009). Developing undergraduate research and inquiry. York: Higher Education Academy, heacademy.ac.uk/assets/York/ documents/resources/publications/DevelopingUndergraduate_Final.pdf, accessed 15 October 2017.

Linking Discipline-Based Research with Teaching to Benefit Student Learning Through Engaging Students in Research and Inquiry, mickhea ley.co.uk/resources, accessed 15 October 2017.

Hopper, L. J., Jr., Schumacher, C. and Stachnik, J. P. (2013). Implementation and assessment of undergraduate experiences in SOAP: an atmospheric science research and education program. *Journal of Geoscience Education*, **61**, 415–427.

Jenkins, A. (2001). How (or whether?) to integrate research into classroom teaching for all students and all higher education institutions. Innovations in Undergraduate Research and Honors Education: Proceedings of the Second Schreyer National Conference 2001, pp. 11–23.

Jenkins, A. and Healey, M. (2012). Research-led or research-based undergraduate curricula. In L. Hunt and D. Chalmers, eds., *University Teaching in Focus: A Learning-Centred Approach*. Victoria, Australia: ACER Press, pp. 128–144.

Kelley, P. H. (1979). *Mollusc Lineages of the Chesapeake Group (Miocene)*. Unpublished PhD thesis, Harvard University.

(1984). Multivariate analysis of evolutionary patterns of seven Miocene Chesapeake Group molluscs. *Journal of Paleontology*, **58**, 1235–1250.

(2004). Time saving through institutional collaboration. *Council on Undergraduate Research Quarterly*, **24**(4), 159.

(2008). Role of bioerosion in taphonomy: effect of predatory drillholes on preservation of mollusc shells. In L. Tapanila and M. Wisshak, eds., *Current Developments in Bioerosion*. Berlin: Springer, pp. 451–470.

(2012). Strategies for teaching evolution in a high-enrollment introductory paleontology course for non-science majors. In M. M. Yacobucci and R. Lockwood, eds., *Teaching Paleontology in the 21st Century*. Paleontological Society, pp. 77–92.

Kelley, P. H. and Dietl, G. P. (2012). The "Research Experiences for Undergraduates in Biodiversity Conservation" program: training the first generation of conservation paleobiologists. *Geological Society of America Abstracts with Programs*, **44**(4), 76.

Kelley, P. H., and Hansen, T. A. (2007). Latitudinal patterns in naticid gastropod predation along the east coast of the United States: a modern baseline for interpreting temporal patterns in the fossil record. In R. G. Bromley, L. A. Buatois, M. G. Mángano, J. F. Genise, and R. N. Melchor, eds., *Sediment-Organism Interactions: A Multifaceted Ichnology*. SEPM Special Publications, **88**, 284–299.

Kelley, P. H., Stanford, S. D., Alexander, C. B., Horne, S. L., Wall, C. N., White, S. M. and Dietl, G. P. (2013a). A low-diversity molluscan assemblage from the Pleistocene of Horry Co., South Carolina. *Geological Society of America Abstracts with Programs*, **45**(7), 327.

Kelley, P. H., Stanford, S. D., Cremer, C. H., Hattori, K. E., Kenison, W., Melcher, L. R., Painter, B. P., Ratchford, R. A. and Dietl, G. P. (2013b). Paleoecological relationships within a molluscan assemblage from the Pleistocene of Horry County, South Carolina. *Geological Society of America Abstracts with Programs*, **45**(7), 321.

Kelley, P. H. and Visaggi, C. C. (2012). Learning paleontology through doing: Integrating an authentic research project into an invertebrate paleontology course. In M. M. Yacobucci and R. Lockwood, eds., *Teaching Paleontology in the 21st Century*. Paleontological Society, pp. 181–197.

Klemm, W. R. (2013). Teaching beginning college students with Adapted Published Research Reports. *The Journal of Effective Teaching*, **13**(2), 6–20.

Koretsky, C. M., Petcovic, H. L. and Rowbotham, K. L. (2012). Teaching environmental geochemistry: an authentic inquiry approach. *Journal of Geoscience Education*, **60**, 311–324.

Kortz, K. M., and Kraft, K. J. (2016). Geoscience education research project: student benefits and effective design of a course-based undergraduate research experience. *Journal of Geoscience Education*, **64**, 24–36.

Leydens, J. A. and Santi, P. (2006). Optimizing faculty use of writing as a learning tool in geoscience education. *Journal of Geoscience Education*, **54**, 491–502.

Lockwood, R., Cohen, P. A., Uhen, M. D., and Ryker, K. A. (2018). Utilizing the Paleobiology Database to provide educational opportunities for undergraduates. In P. Cohen, L. Park Boush, and R. Lockwood, eds., *Pedagogy and Technology in the Modern Paleontology Classroom. Elements of Paleobiology*, **1**, **-**.

Lopatto, D. (2010). *Science in Solution: The Impact of Undergraduate Research on Student Learning*. Washington, D.C.: Council on Undergraduate Research.

Mayborn, K. R. and Lesher, C. E. (2000). Teaching the scientific method using contemporary research topics as the basis for student-defined projects. *Journal of Geoscience Education*, **48**, 145–149.

Montgomery, H. and Donaldson, K. (2014). Using problem-based learning to deliver a more authentic experience in paleontology. *Journal of Geoscience Education*, **62**, 714–724.

National Academies of Sciences, Engineering, and Medicine. (2017). *Undergraduate Research Experiences for STEM Students: Successes, Challenges, and Opportunities*. Washington, DC: The National Academies Press. doi:https://doi.org/10.17226/24622.

National Science Board. (2011). National Science Foundation's Merit Review Criteria: Review and Revisions, nsf.gov/nsb/publications/2011/meritre viewcriteria.pdf, accessed 15 October 2017.

National Science Foundation. (2014). Investing in Science, Engineering, and Education for the Nation's Future: Strategic Plan for 2014–2018, nsf.gov /pubs/2014/nsf14043/nsf14043.pdf, accessed 15 October 2017.

National Science Foundation. (2017). Report of the 2017 Committee of Visitors, Division of Earth Sciences, Directorate for Geosciences, National Science Foundation, 2014–2016 Review Period, nsf.gov/geo/adgeo/advcomm/ fy2017_cov/ear-cov-2017-report.pdf, accessed 31 October 2017.

National Society for Experiential Education. (1998). Eight Principles of Good Practice for All Experiential Learning Activities. NSEE Annual Meeting, Norfolk, VA. Updated December 2013, nsee.org/8-principles, accessed 15 October 2017.

Neogene Marine Biota of Tropical America. (2016), porites.geology.uiowa.edu/, accessed 15 October 2017.

Paleobiology Database. (2017). paleobiodb.org/#/, accessed 15 October 2017.

Peterson, C. D., Anderson, L. L. and Michtom, W. D. (1996). Applications of undergraduate research proposals in general-education earth-science courses. *Journal of Geoscience Education*, **44**, 197–201.

Robinson, G. D. (1987). Using journal articles in an introductory geology class. *Journal of Geological Education*, **35**, 140–142.

Science Education Resource Center. (2016). Integrating Research into Geoscience Courses, https://serc.carleton.edu/NAGTWorkshops/careerprep/teaching/ IntegratingResearch.html, accessed 15 October 2017.

Tessier, J. T. (2012). Effect of peer evaluation format on student engagement in a group project. *The Journal of Effective Teaching*, **12**(2), 15–22.

University of North Carolina Wilmington. (2017). 2017–2018 Undergraduate Catalog, catalogue.uncw.edu, accessed 16 October 2017.

Visaggi, C. C., Dunham, J. I., Griffin, C. T., Kirkland, J. C., Rossi, D. V., Thiery, D., Pickering, R. A., Parnell, B. A., Kelley, P. H. and Dietl, G. P. (2014). Drilling predation in molluscan assemblages of the Lower Waccamaw Formation (Pleistocene) at Snake Island Pit in North Carolina. *Geological Society of America Abstracts with Programs*, **46**(3), 94.

Wei, C. A. and Woodin, T. (2011). Undergraduate research experiences in biology: alternatives to the apprenticeship model. *CBE Life Sciences Education*, **10**, 123–131.

Wenzel, T. J. (1997). What is undergraduate research? *Council on Undergraduate Research Quarterly*, **17**, 163.

Acknowledgments

I thank R. Lockwood, P. Cohen, and L. Park Boush for organizing this short course and inviting me to participate. Discussions with T. Hansen initially spurred me to incorporate authentic research in my courses. The research conducted by the invertebrate paleontology cohorts focused on here was an extension of work supported by National Science Foundation Grant No. EAR-0755109; G. Dietl, J. Smith, S. Durham, and S. Neely were involved in collecting samples discussed here. C. Visaggi, S. Stanford, and S. Neely served as invertebrate paleontology teaching assistants during 2007–2008, 2011–2013, and 2015 respectively and were involved in mentoring the students. The University of North Carolina Wilmington ETEAL initiative funded the participation of S. Stanford in 2013. K. Bruce co-taught "Behavior of Animals: Dead and Alive" with me and, as director of the UNCW Center for Support of Undergraduate Research and Fellowships, provided funds for undergraduate travel to GSA meetings. J. Register, S. Whittemore, and M. Kirby provided access to collecting sites for the classes discussed here. P. Fritzler and W. Wilkinson conducted training workshops for students in library use and technical writing. My thanks to the 99 students who took invertebrate paleontology with me from 2003 to 2015, and especially to the 10 students from the 2013 cohort who counted the 23,276 *Mulinia*.

Cambridge Elements ☰

Elements of Paleontology

Editor-in-Chief

Colin D. Sumrall
University of Tennessee

About the Series

The Elements of Paleontology series is a publishing collaboration between the Paleontological Society and Cambridge University Press. The series covers the full spectrum of topics in paleontology and paleobiology, and related topics in the Earth and life sciences of interest to students and researchers of paleontology.

The Paleontological Society is an international nonprofit organization devoted exclusively to the science of paleontology: invertebrate and vertebrate paleontology, micropaleontology, and paleobotany. The Society's mission is to advance the study of the fossil record through scientific research, education, and advocacy. Its vision is to be a leading global advocate for understanding life's history and evolution. The Society has several membership categories, including regular, amateur/avocational, student, and retired. Members, representing some 40 countries, include professional paleontologists, academicians, science editors, Earth science teachers, museum specialists, undergraduate and graduate students, postdoctoral scholars, and amateur/avocational paleontologists.

Paleontological
S O C I E T Y

Cambridge Elements $^{\equiv}$

Elements of Paleontology

Elements in the Series

These Elements are contributions to the Paleontological Short Course on *Pedagogy and Technology in the Modern Paleontology Classroom* (organized by Phoebe A. Cohen, Rowan Lockwood, and Lisa Boush), convened at the Geological Society of America Annual Meeting in November 2018 (Indianapolis, Indiana USA).

A full series listing is available at: www.cambridge.org/EPLY

For EU product safety concerns, contact us at Calle de José Abascal, 56–1°,
28003 Madrid, Spain or eugpsr@cambridge.org.

www.ingramcontent.com/pod-product-compliance
Ingram Content Group UK Ltd.
Pitfield, Milton Keynes, MK11 3LW, UK
UKHW050729090126
466816UK00013B/254